I0038114

# CHANGING TRENDS IN COSMETICS PACKAGING

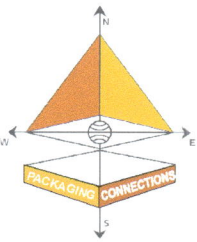

Sanex Packaging Connections
Pvt. Ltd.
www.packagingconnections.

# Copyright

**Published by :**

**Sanex Packaging Connections Pvt. Ltd.**

An ISO 9001 : 2008 Certified Organisation

117, Suncity Tower, Sector-54

Golf Course Road, Gurgoan-122 002.

**Tel : +91 124 4965770**

**Fax : + 91 124 41433951**

**e-mail : info@packagingconnections.com**

**Like us on Facebook : www.facebook.com/pconnection**

**ISBN :** 978-8192792019

© **Reserved with the publisher**

All Right Reserved. No part of this publication may be reproduced or transmitted in any form or by any means, electronic or mechanical, including photocopying, recording or any information storage and retrieval system, without permission in writing from the publisher.

All trademarks and images that appear or are related to the artwork featured in this book belong to their respective artists and/or companies that commissioned the work and/ or feature herein.

**CHANGING TRENDS IN COSMETICS PACKAGING**

# List of Contributors

Team www.PackagingConnections.com by Sanex Packaging Connections Pvt Ltd

Sandeep Kumar Goyal, Founder & CEO
Amita Venkatesh Vallesha, Associate: Scientific Affairs & Consultancy
Chhavi Goel, Associate: Research & Business Consulting
Bhaskar Ch, Technology Advisor e-business
Sonu Sheoran, Associate Research & Technology
Ashok Kumar, Programme Manager: KPO

# Table of Contents

# Introduction

This publication comes after the success of Ideas & Opportunities 2013 held on 19th July 2013 in India. Various innovations were presented during the one day workshop by the expert consulting team of Sanex packaging Connections Pvt Ltd popularly known as Team PackagingConnections.

Idea behind this is to bring the innovations to wider group of professionals to meet the mission of packaging knowledge sharing and that too cost effectively. We feel that this publication will further fill the project pipelines of companies and improve the standards of packaging. Many professionals either do not have the access or time to go through so many innovations together. So we think this publication will fill that gap. For your feedback please email directly to info@packagingconnections.com

With this, Enjoy Wonders Of Packaging!

**Sandeep Kumar Goyal**
Founder & CEO ,
www.PackagingConnections.com

# COSMETICS DESIGN

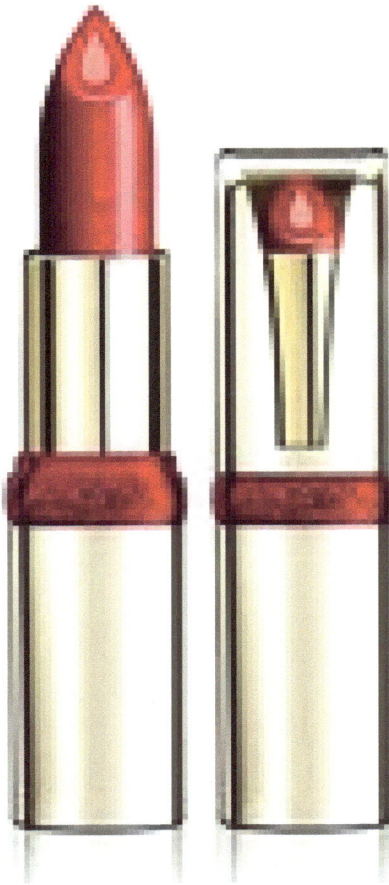

## Manufacturer/Designer

**Designed for Loreal
Manufactured by Rexam
personal care**

- A designer display window to showcase the product.

- This is the specific case for anti-ageing serum mix lipstick

- The window allows immediate visualization of the special product

- High precision laser technology to create crystal clear window area

## Manufacturer/Designer

**Brainbox Design**
**For Capricho brand**
**M/f by Boticario, Brazil**

- Innovative carton packaging design for combo pack of deodorant.
- The colour scheme shows the day and night usage of the products.
- Attractive shelf pack

**CHANGING TRENDS IN COSMETICS PACKAGING**

## Manufacturer/Designer

**Eva Karpati**
**Karpati Corporation Pty Ltd**
**Australia + 61 2 9009 6666**
**www.karpati.com.au info@karpati.**
**com.au**

- A unique design that offers real consumer convenience.

- The bottle slips over the fingers like a ring, no flat surface needed to steady the bottle.

- Bottle is of Carbonated Polypropylene material manufactured by Baralan International.

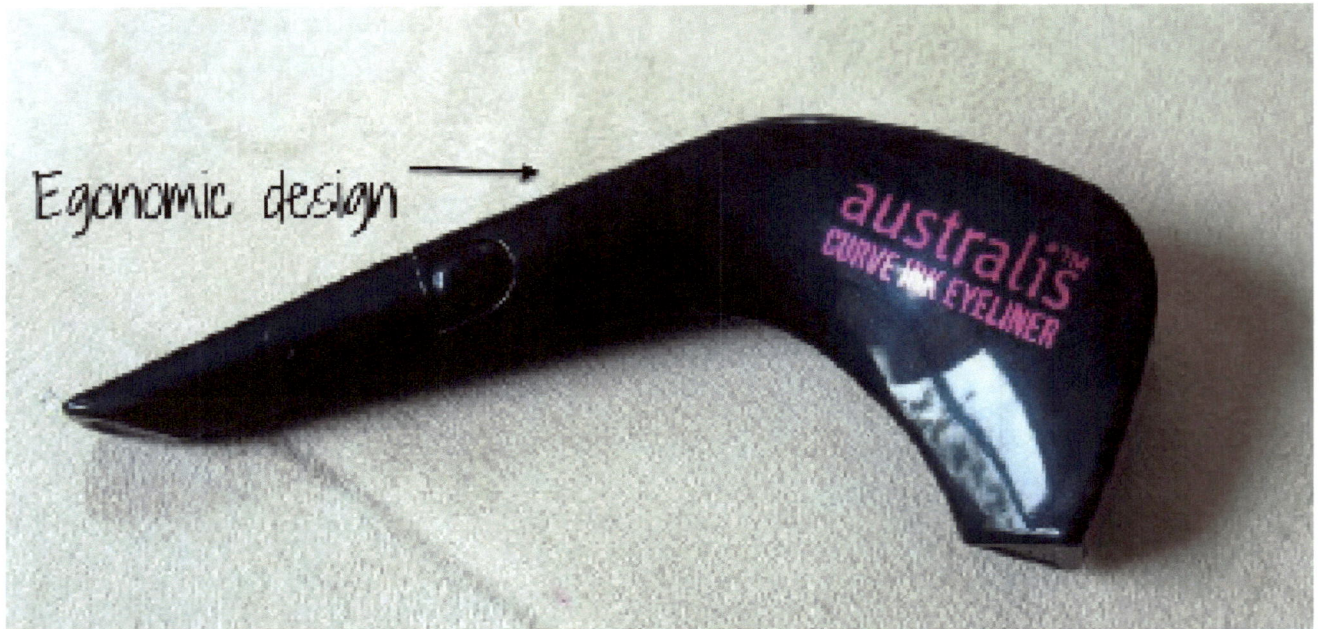

Egonomic design →

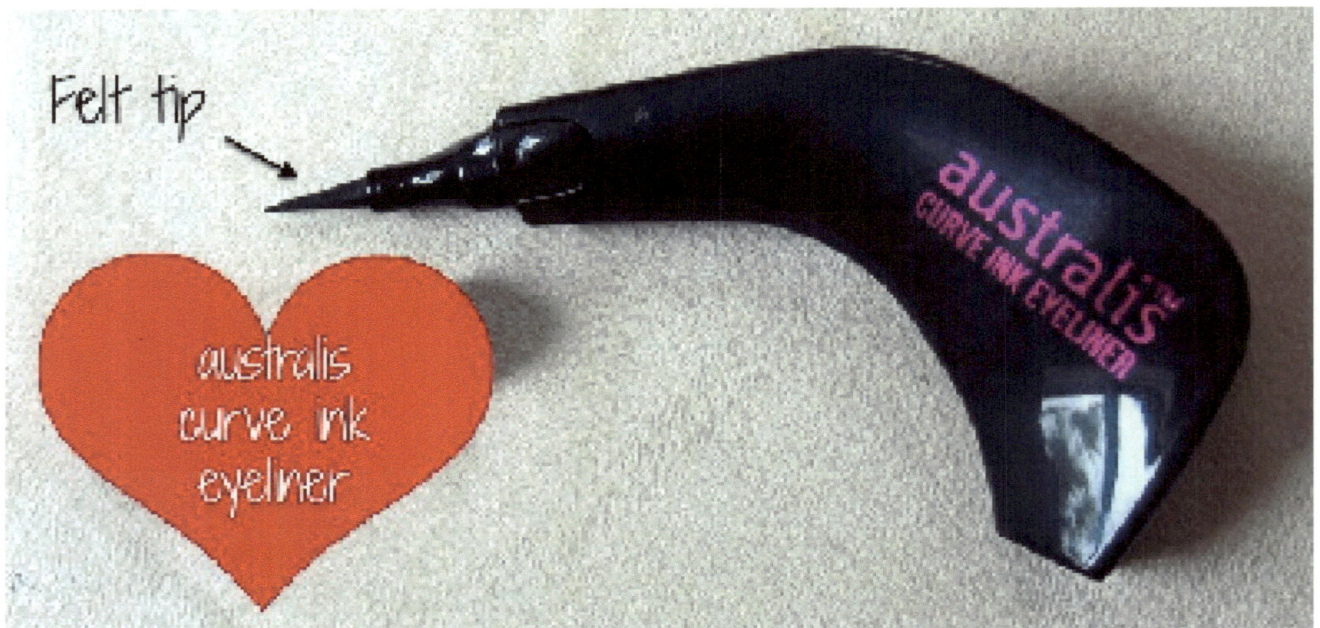

Felt tip

australis
curve ink
eyeliner

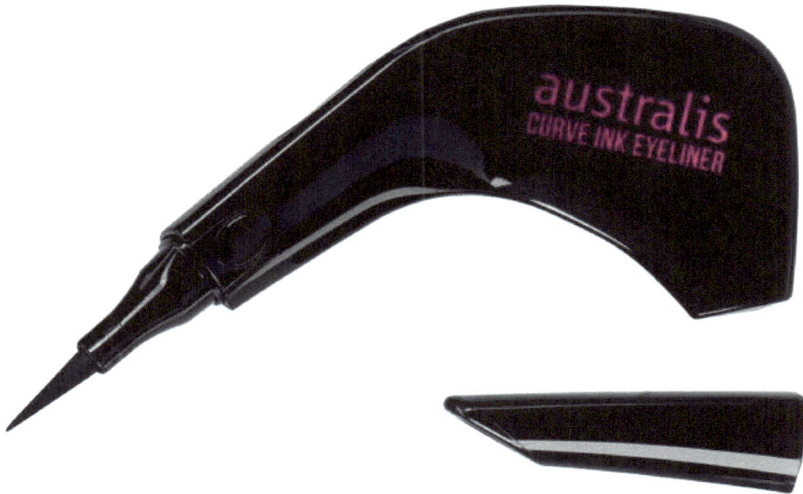

## Manufacturer/Designer

**Designed for Australis**
**Cosmetics**
**1300650981**
**www.australiscosmetics.com.au**

- The curve shapes provides ease in application

- The applicator brush has a multifunctional tip — fine brush to give thin line, if pressed down gives thicker line

- Specifically designed to be used by left hand as well that provides a grip zone

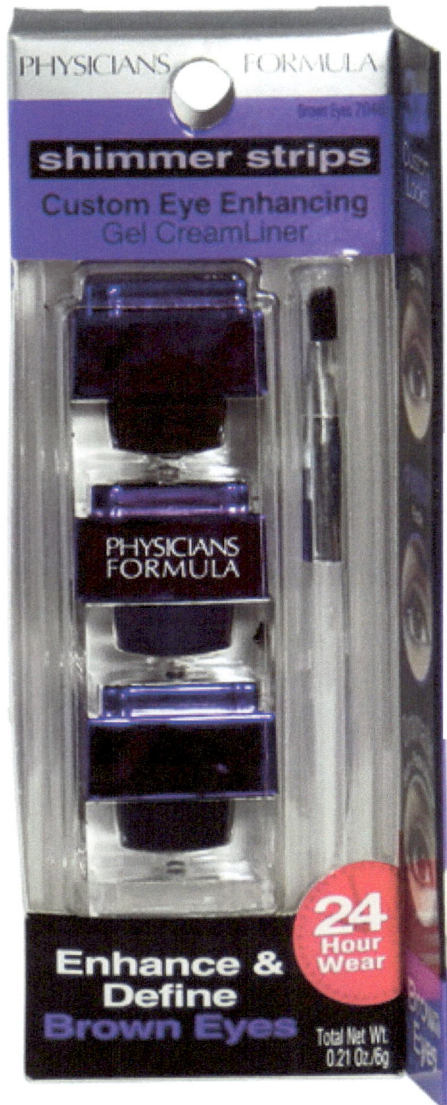

## Manufacturer/Designer

**Designed for Physicians Formula**
**http://www.physiciansformula.com**

- The design gives an option for enhanced branding when products are stacked on shelf

- This design can be well utilized for combo packs of cosmetic products — multiple shades of eye shadow, lipsticks and eye liners

- Aligning position of different unit packs, multi directional branding can be achieved

## Manufacturer/Designer

**Developed by Measa Engineering (NY, LA, Paris) for Benefit Cosmetics**
**http://www.maesa.com/**

- The bottles have a grooved texture and the surface under the labels is slightly depressed

- The product is differentiated on the shelf as it not only gives attractive look but also gives a unique feel when touched

- The cap is given a texture that mimics the "cork" look

- The cap is dipped and coated with a special film, creating the distinct effect.

## Manufacturer/Designer

**Developed by Arrowpack for Physicians Formula**
**www.arrowpack.com**

- Small sized bottles generally for combo packs of 3-4 nail paints
- The base of the upper bottles fits snugly into the well of the lower lids.
- This also gives attractive shelf appeal for nail polish brands that generally occupy queued horizontal space on the shelves

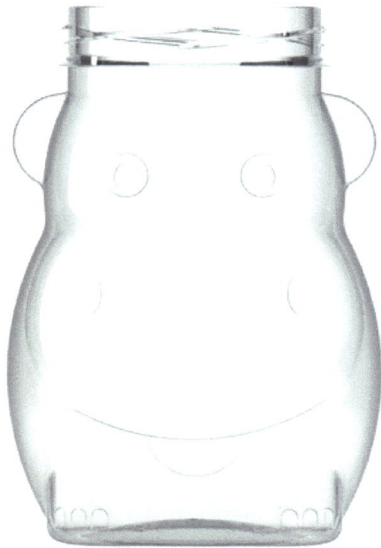

## Manufacturer/Designer

**Developed by PET Power**
**Marcel Schröder (MD) /**
**Kees Kok (Marketing)**
**+31 (0)76 503 82 83**
**http://www.petpower.eu**
**kk@petpower.eu**

- Generaly we have seen such profile bottles made of HDPE or PP

- Made out of PET these profile shaped bottles and jars can be used for packaging of products targeted for Kids like shampoo, oil, lotions, creams etc

## Manufacturer/Designer

### Heinz Glas Klaus

- This is an example of unique design that can be obtained from a high quality decoration glass.

- The bottle can be used for perfumes, premium female cosmetic range like face creams, body lotions, deodorants

- The pack gives high shelf appeal and the customized design immediately reflects it's target consumer

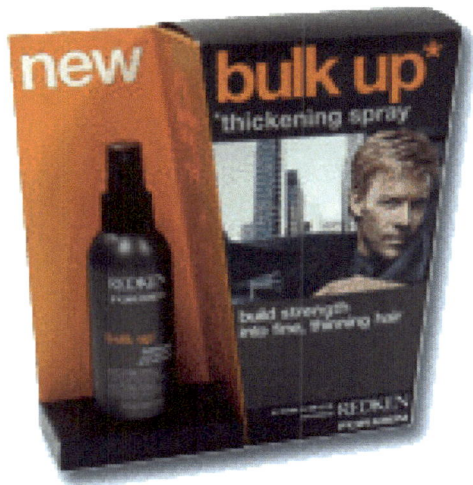

## Manufacturer/Designer

**Diamon Packaging (NY)**
**585-334-8030 / 800-333-4079**
**http://www.diamondpackaging.com**
**sales@diamondpkg.com**

- The geometrically-inspired design combines a trapezoid shape.

- The inverted weight of the main panel contributes by visually communicating the sense of "bulking up".

- The fifth-panel structure integrates with the base on which the spray bottle is displayed, encouraging consumers to touch and hold.

- This pack gives an innovative POP idea for the versatile range of products

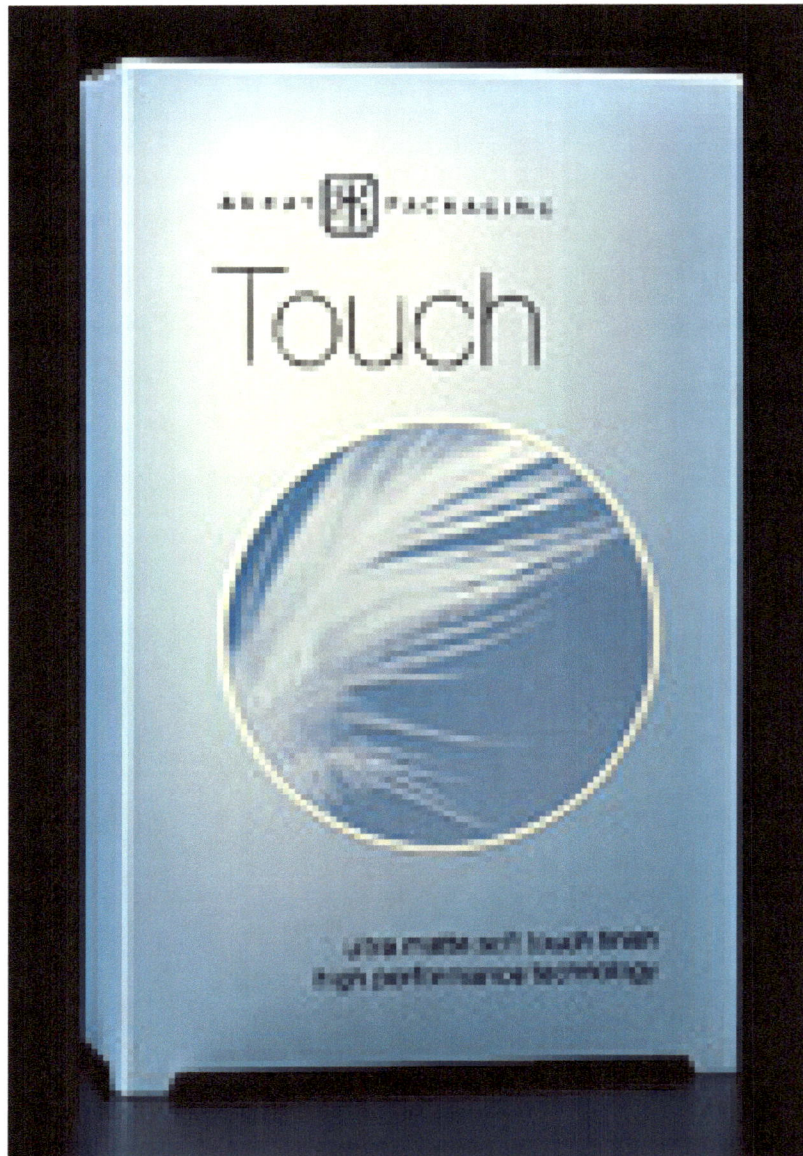

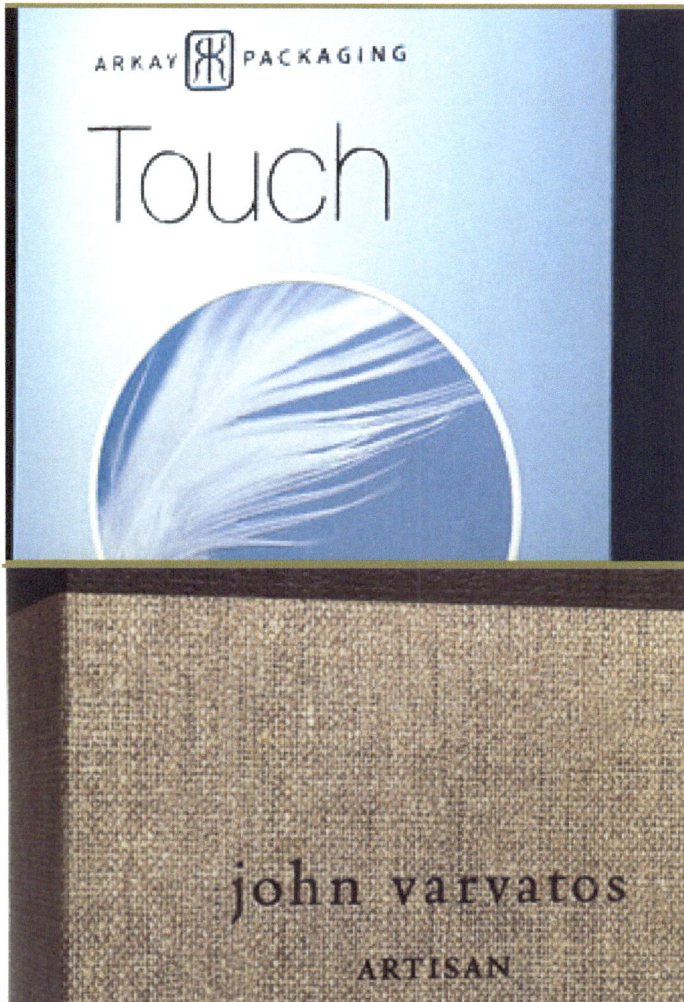

## Manufacturer/Designer

**Arkay Packaging**
**Roanoke, Virginia 540.977.3031**
**roanoke@arkay.com**
**http://www.arkay.com**

- These are high performance aqueous coating that produces a finish, which is soft to the touch.

- This can be used for packaging of premium cosmetics specially face creams that relate to the "soft feel" given by the coating.

- Consumer can feel the special effect from the outer carton itself since the coating acts to give marketing leverage to the pack

**CHANGING TRENDS IN COSMETICS PACKAGING**

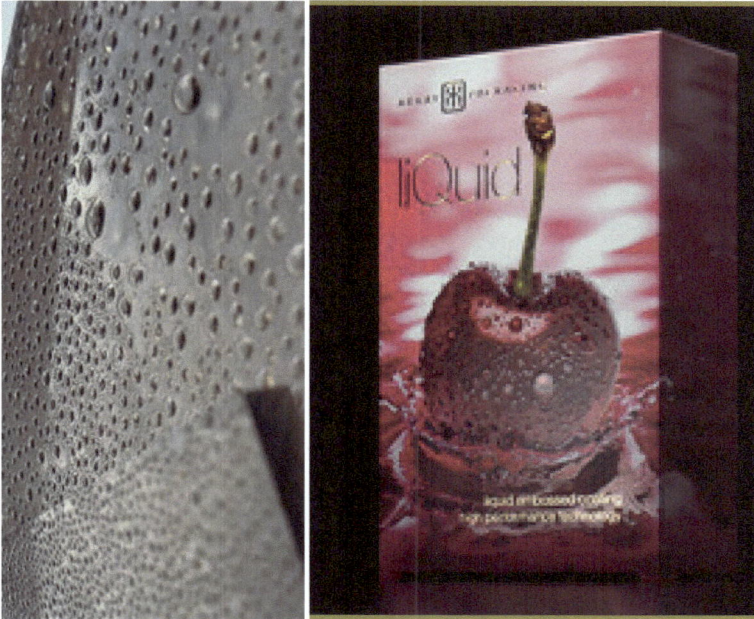

## Manufacturer/Designer

**Aarkay Packaging**
**Roanoke, Virginia**
**540.977.3031**
**roanoke@arkay.com**
**http://www.arkay.com**

- A specialist UV Cured coating which is formulated using two opposing lacquers.

- The lacquers are over coated on one another and once cured, the coating reticulates and forms a bead or droplet.

- It will not fade, crack or block under normal usage conditions and will not interfere with cello wrap and/or heat tunnel applications or final packaging.

- The coating requires no post set-up time, is available for final processing immediately after being applied, and won't impact production scheduling.

- In addition, this is environmentally friendly, consisting of recyclable material.

## Manufacturer/Designer

**Bottles from M&H Plastics**
**www.mhplastics.com**
**Labels by CCL Label**
**www.ccllabel.co.uk**

- The idea behind the animal label concept is to impart the message of product's natural organic and animal friendly ingredients.

- However the idea also gives an attractive look to the packaging.

- Currently used for Hair-care products

- Bottles are made of HDPE with flip top caps

- 360 degree pressure sensitive labels for 100ml sku and decorative shrink sleeves for 250ml sku

# COSMETICS
# BREAK THROUGH

# Soft-Tips Applicator

Pinpoint

Ribbontip

**CHANGING TRENDS IN COSMETICS PACKAGING**

## Manufacturer/Designer

**Developed by Aptar**
**www.aptar.com**
**Available for tubes only from**
**Albea, CCL, Berry, Neopac, Linhardt**

- The tube has soft tip (made of silicon or EPDM)

- Two shapes are available — Pin-point and Ribbontip

- Beneficial for precise and controlled dispensing

- Always give smooth, clean touch tip during application

- The target end uses are — dark circle treatment, nail care, concealers, acne treatment.

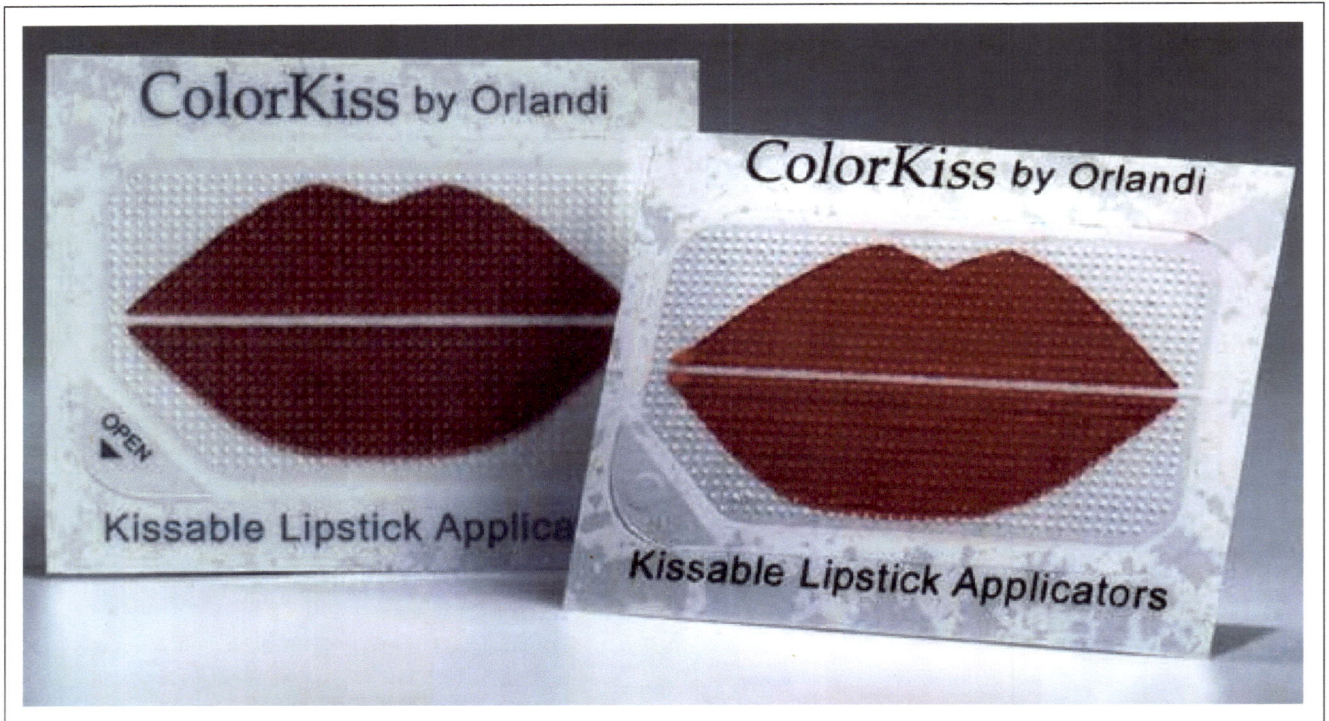

ColorKiss by Orlandi
OPEN
Kissable Lipstick Applica

ColorKiss by Orlandi
Kissable Lipstick Applicators

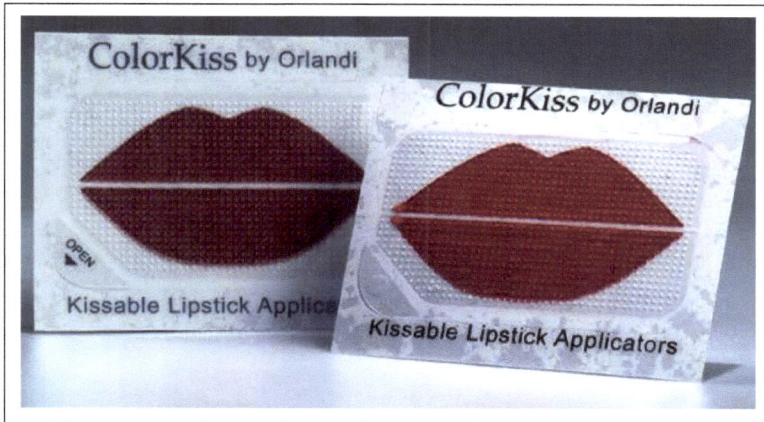

## Manufacturer/Designer

Orlandi Inc, US
0631-756-0110
http://www.orlandi-usa.com
fritts@orlandi-usa.com;
info@orlandi-usa.com

- 8 mil thick substrate of paper (bleached) laminated with PET on both sides using ethylene copolymer
- This is treated before applying lipstick (technology by Orlandi)
- Covered with peelable PET film (3 mil) by heat sealing
- Provided by embossed base layer that keep lipstick to be in contact with the top layer
- It is then bent backward and slided on lips to apply
- It is smaller than a business card
- ColorKiss® is manufactured under U.S. and foreign patent

# DUO AIRLESS PACK

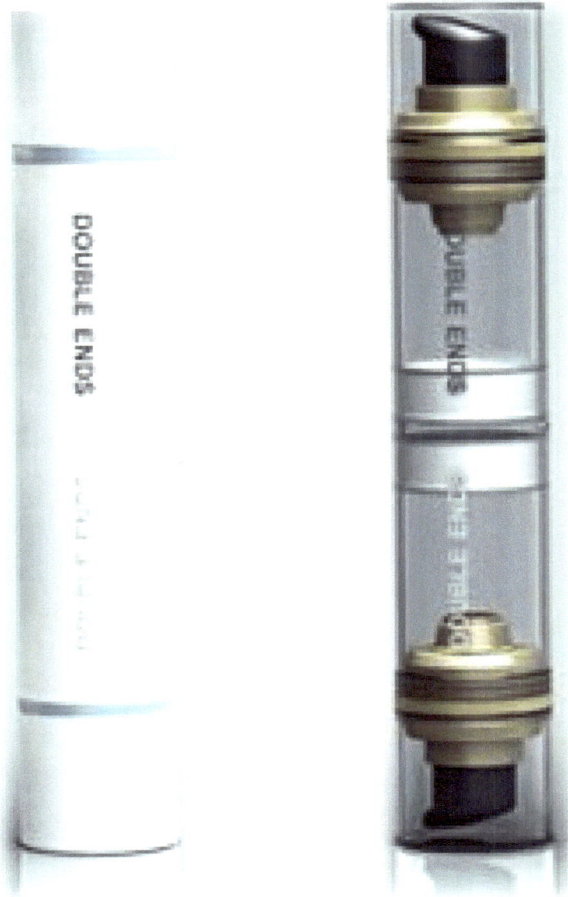

**CHANGING TRENDS IN COSMETICS PACKAGING**

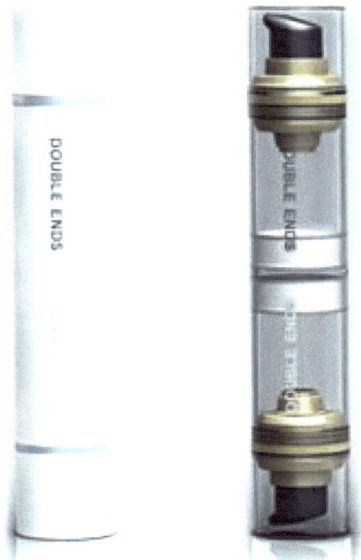

## Manufacturer/Designer

**Quadpack Grp./Yonwoo**
**032-575-8811; Korea**
**http://www.quadpack.net/uk/**
**our-products/featured-products/**
**travelling-light--yonwoo-s-duo-**
**airless-two-in-one/ webmaster@**
**yonwookorea.com**

- This is a double-ended airless bottle designed for travel-sized two-part treatments.

- No need to carry two products around

- Ideal for day and night cream, eye and face cream, primer and foundation

- Technical specifications include 15ml capacity each

- PP material in contact with the formula, an airless pump on both ends and easy top filling.

## Manufacturer/Designer

**Material from EMX Grivory**
**Manufactured using Ganhal IBM**
**http://www.emsgrivory.com/en/**
**http://www.ganahl.ch/en/**

- Manufactured using transparent amorphous Polyamide

- Target applications may be nail paint bottles, perfume bottles.

- Glass is breakable with good barrier, plastics have poor barrier but are non breakable.

- This material gives both unbreakable and high barrier properties

- Due to high barrier it prevents the evaporation of volatile ingredients, subsequently improving products durability

## Manufacturer/Designer

**Developed by Quadpack group**
**http://www.quadpack.net**

- Stick packs are a fast growing solution for single-use pharmaceutical applications, providing lightweight, secure and hygienic packaging, perfect for travelling and easy disposal

- Good compatibility between the productand package

- Excellent seal integrity for highly sensitive or aggressive drugs

- Easy to squeeze, convenient andhygienic dispensing

- Minimal waste of product and packaging

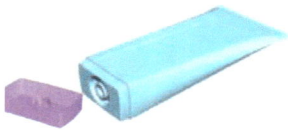

Rectangle, integrated dispensing valve,
embossed sqeeze-area

Triangle, integrated disc-top closure, embossing

**Clean**

Oval, integrated flip-top closure,
embossed sqeeze-area

Round, integrated disc-top closure

## Manufacturer/Designer

**Developed by Plasticum Group**
**www.plasticumgroup.com**
**info@plasticumgroup.com**

- The Tubes are manufacured by injection moulding, using a unique patented material formulation and tooling techniques

- Customized shapes possible (triangular, square, etc)

- Integrated dispensing valve

- Integrated closure

- In-Mould-Labelling, embossing, textured surfaces

- The name given is CLUBE® = Closure + Tube

VS

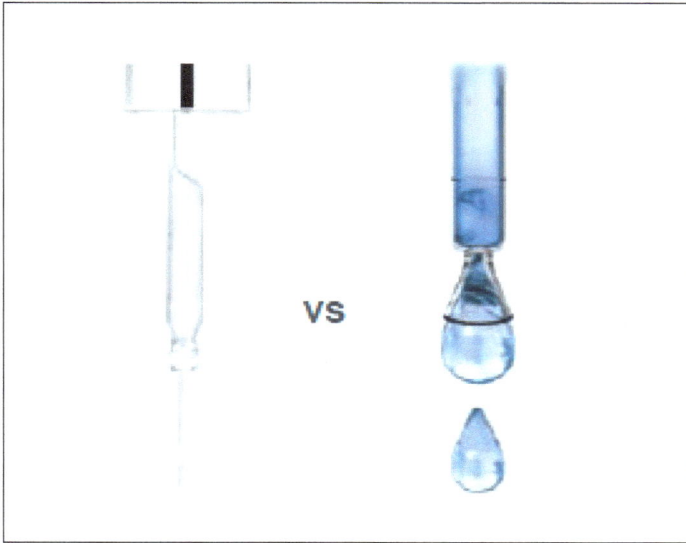

## Manufacturer/Designer

**Eun Sub Lee (Director)**
**Korea Industries Inc.**
**82-2-9002868**
**www.fskorea.com**
**sudlder@fskorea.com**

- Funnel shaped pipette end dropper
- Provides perfect dosage compared to normal droppers
- Great for spot care treatments
- Effective for accurate dosage at areas like eye rim, blemish spots

# Metal Tipped Mushroom Cap

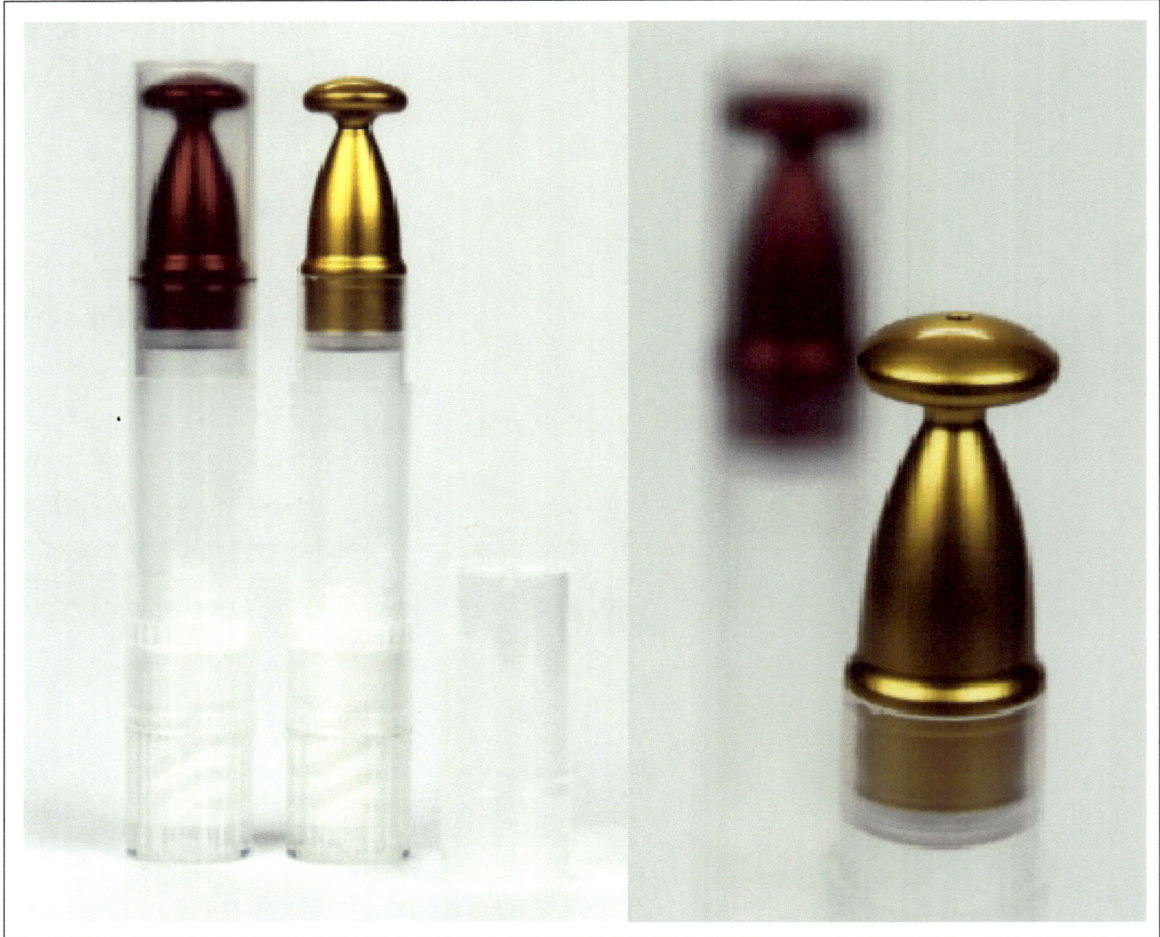

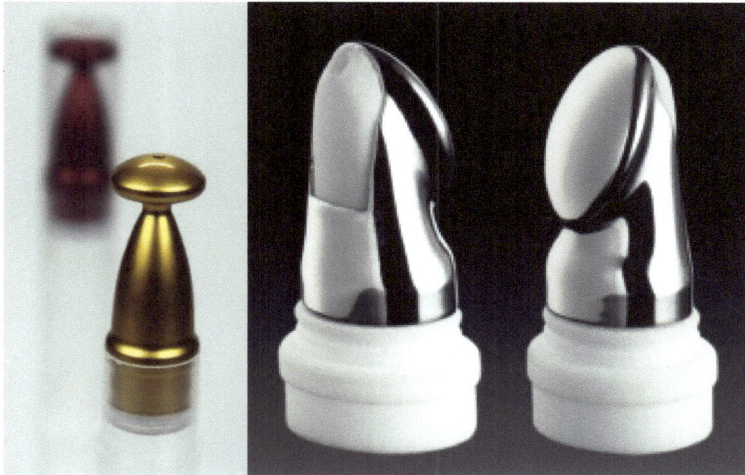

## Manufacturer/Designer

**Chenyuan Lin / Gina Lin**
**Libo Cosmetics**
**+ 86 769-22675858 /**
**22089761/ 23118009**
**http://www.libocosmetics.com**
**sales@libocosmetics.com**

- Unique shaped head - much like a mushroom cap.

- Perfect for the hands-free application of wrinkle treatments around the eyes and mouth where touching the formula isn't recommended

- The product reservoir, made of a PP interior with a ratcheting SAN base, allows consumers to twist the bottom to dispense minuscule amounts of product out of the top aperture to be spread with the metal head.

# VALUE
# ENGINEERING

## Manufacturer/Designer

**Dieter Bakic Design GmbH**
**München, Germany**
**+49-89-490 436-0**
**marketing@bakic.com**

- We have seen "disc top" closures but they have their own constraints.
- This is an innovative cap that has a new opening and closing mechanism.
- When the cap is turned, a disc lifts or descends, opening and closing the dispensing hole built into the cap.

**CHANGING TRENDS IN COSMETICS PACKAGING**

## Manufacturer/Designer

**Designed by Questto No**
**For SOU brand**
**M/f by Natura, Brazil**

- Replacement of rigid containers with flexible stand up pouch
- Unique shape that gives shelf appeal
- 75% less plastic used then the conventional pack
- Less space while transportation, storage.

**CHANGING TRENDS IN COSMETICS PACKAGING**

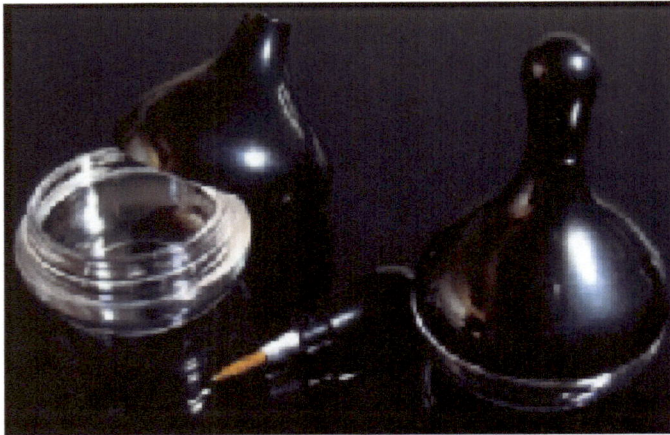
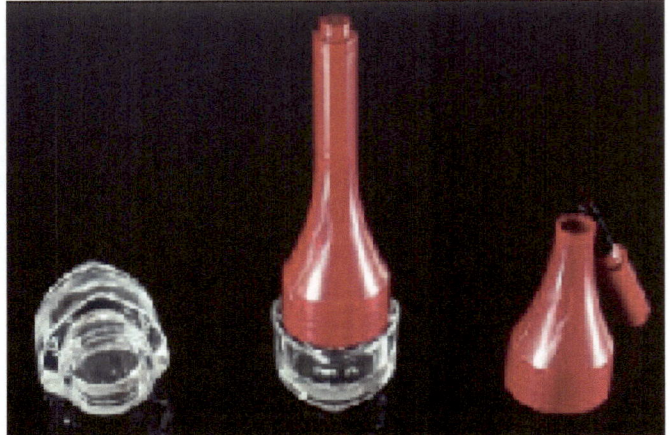

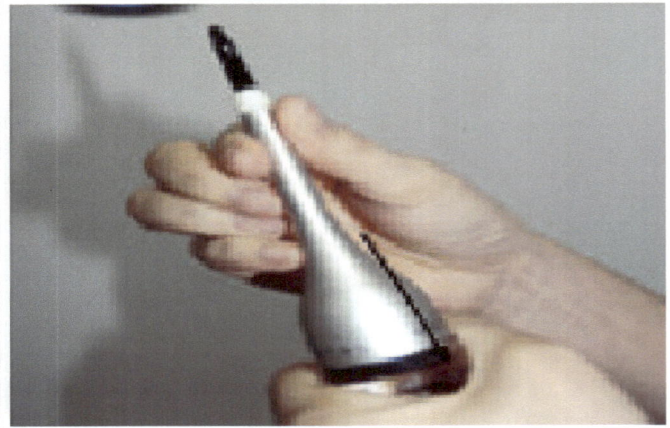

**CHANGING TRENDS IN COSMETICS PACKAGING**

## Manufacturer/Designer

Marc Tsao (Sales)
Fancy & Trend, Taiwan
886-2-29874374
http://www.fancyandtrend.com.tw
sales@fancyandtrend.com.tw

- Base material is made of SAN (compatible with ingredients for long lasting and non smudge formulation)

- Top cap and brush of ABS (ease in decoration / metalizing / over spray)

- Counterparts have 2 part system while this is 2+1

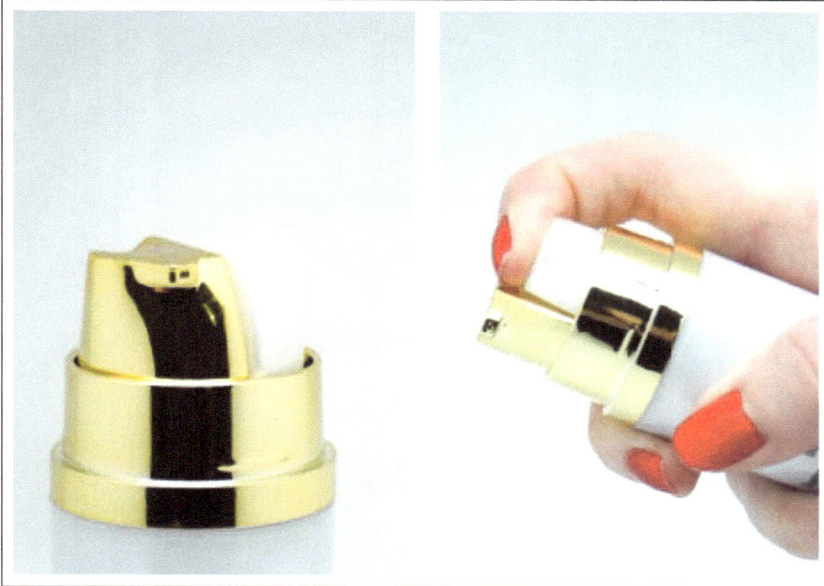

## Manufacturer/Designer

Allen Chen
Gidea Packaging Co. Ltd.
86-574-27865689 / 137 0574 478
http://www.gideapackaging.com/
allenchen@gidea.cn

- For multi-phase solutions that cannot be mixed prior to using.

- Container is split into two distinct reservoirs with complementary dispensing pumps for individual product

- Lesser material usage

- Made of ABS with clear PP lining

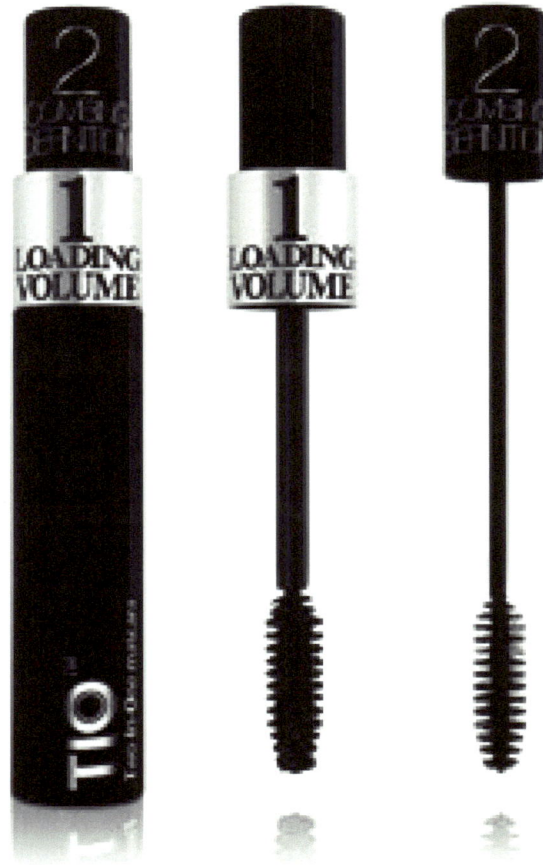

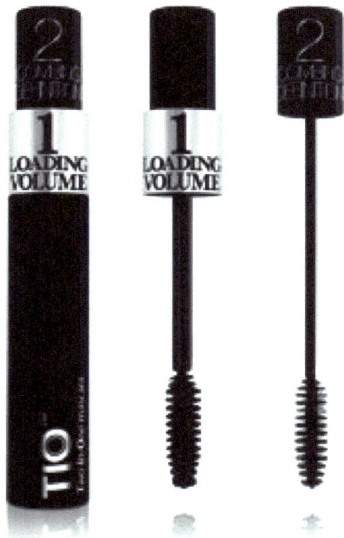

## Manufacturer/Designer

**Developed by Albea Group**
**www.albea-group.com**

- TIPTM — Two in one mascara that serves to give multi effects.

- Normal opening with a normal brush that gives heavy loading of the product.

- Additional opening through the first brush that gives light loading to be used for under eye lashes or for light makeup (during day time).

## Manufacturer/Designer

**Developed by Albea Group**
**www.albea-group.com**

- The tube is named as "wellness brush tube".

- Scrubbing and stimulating brush applicator on a big volume tube.

- Ideal for foot treatments

- No battery operated mechanism.

- The simple technology involves incorporation of a dispensing cap with inbuilt brushes.

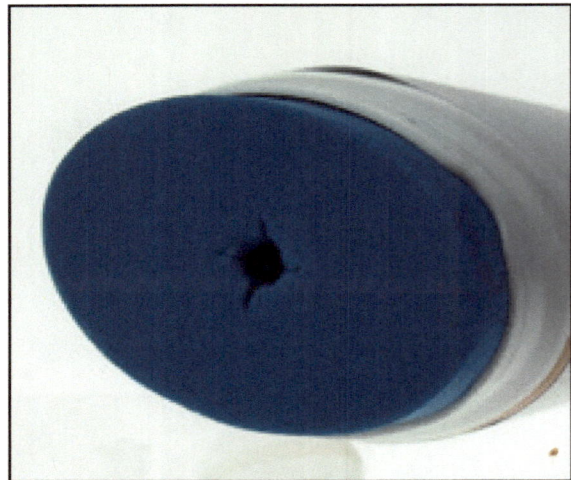

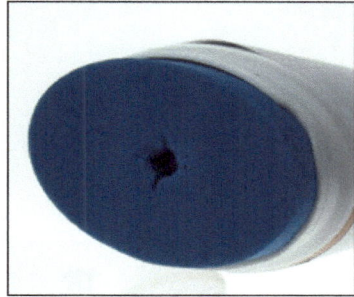

## Manufacturer/Designer

**Developed by Quadpack Group**
**http://www.quadpack.net**

- With its innovative dispensing method and soft foam head, the Hands Free tube allows easy and gentle application.

- Material used is PE ad PP

- Orifice shape and size can be customized based on viscosity of the product to dispense

- End use targets are hair remover creams, insect repellent creams, pain relief lotion, sunscreen, etc

**CHANGING TRENDS IN COSMETICS PACKAGING**

10,000+
POUNDS
LESS
PLASTIC
PER ONE
MILLION
TUBES

95% dedicated to branding

Savings

10,000 LBS.
LESS PLASTIC
USED PER
ONE MILLION
TUBES

## Manufacturer/Designer

**Developed by Tricor Braun**
**US / Canada**
**+1-314-569-3633**
**www.tricorbraun.com**

- These are about 50% lighter than traditional closure and tube heads, thus uses 10,000 pounds less plastic per one million tubes.

- Benefits —

  — fewer trucks on the road and a smaller carbon footprint

  — Cost Effective

  — Optimum Performance

  — Brand Optimization - The package's minimalist design creates even more space for dedicated branding

  — Customization

- The new, light weight closure and low profile tube head will be commercialized during the fourth quarter 2013.

# Monodose Sachet

NEW

Double Serum
[Hydric + Lipidic System]

CLARINS
PARIS

Double Serum

**CHANGING TRENDS IN COSMETICS PACKAGING**

bend  snap  squeeze

## Manufacturer/Designer

**Developed by Easysnap**
**(+39) 051.68.10.804**
**For the Clarin's brand**
**http://www.easysnap.com info@**
**easypacksolutions.com**

- A new variation in a well known "monodose sachet" — Bidose

- Launched for the double serum cream for intensive anti-aging treatment

- Provides required shelf life and can pack any type of liquid or viscous product

- Bi-dose sachet can be used for packaging of cosmetics or serums having separate products that cannot be mixed prior to their usage

- Single time use

# CONCEPTS

# POP Pack for Sachets

## Manufacturer/Designer

**Developed by group of students**
**Shweta Budukh**
**+91-9011026976**
**shweta.budukh@gmail.com**

- The display pack is made up of acrylic for see through and aesthetic appeal.

- The pack is provided with partition; to allow multi-variant display and multi-faced branding

- In every partition rod is provided for hanging in a way that so sachets can be pulled and tear off easily

- To dispense sachets from opposite sides and adjacent sides of the pack, it can be rotated with ball bearing mechanism fixed at the bottom

- Side Labels can be used to identify the product variant and according to that pack can be rotated.

- This display pack provides excellent branding

- Made of rigid plastic hence can be reused with minimum wear and tear

**CHANGING TRENDS IN COSMETICS PACKAGING**

## Manufacturer/Designer

**Designed by Veronica Clauss**
**United Kingdom**
**+1 570 352 2009**
**http://www.veronicaclauss.com/**
**toothpaste.html**
**veronicaclauss@gmail.com**

- The pack is designed for combo of morning and evening toothpaste packaging
- The design is such that it gives 3-D look to the product.
- Provides better shelf appeal than a traditional rectangular pack.
- Can be ideated for other products as well. (pharma creams, ointments, day and night creams, etc)

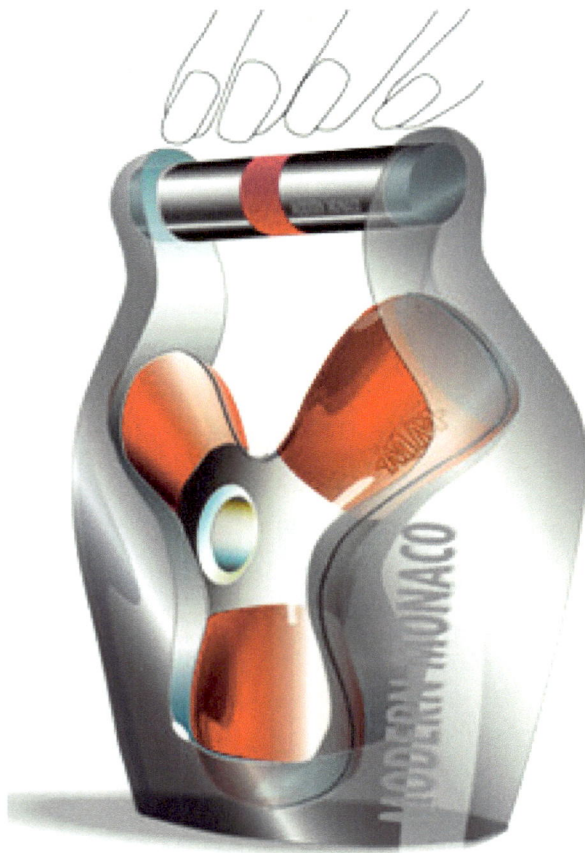

Mirror Housing

Powder Housing

Applicator Housir

**CHANGING TRENDS IN COSMETICS PACKAGING**

## Manufacturer/Designer

### Designed by Fernando

- This is a retail design for a powder compact, mirror, applicator and a lipstick

- The three main elements in the shape 'Y'.

- The final design consists of two separate shells, which house the inner components.

- The hollow centre of the spindle allows for the integration of the lipstick housing.

- The packaging is made of a frosted rubber material molded in the form of the letter U, holding together the pieces to this product.

- The horizontally placed lipstick tube serves as a carry handle for the package.

**CHANGING TRENDS IN COSMETICS PACKAGING**

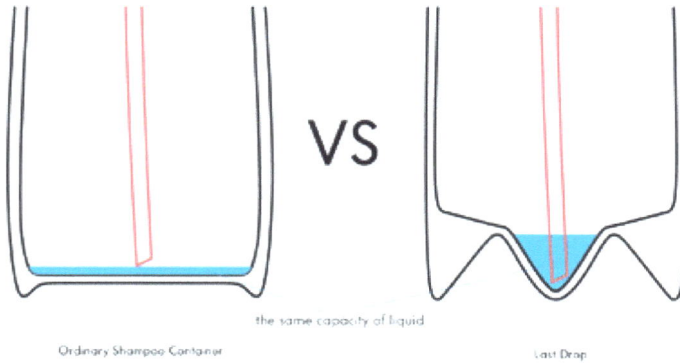

VS

the same capacity of liquid

Ordinary Shampoo Container

Last Drop

## Manufacturer/Designer

**Developed by team of students from Samsung Art & Design Institute (SADI) - Seonkeun Park and Jinsun Park**

- The concept is derived from the inconvenience of pumping up the finishing liquid from the bottom of a container to avoid everyday wastage.

- The bottom is designed in a way that the finishing liquids are stored in a cone following by slopes both side where the dispenser can easily reach and let the user to have even the last drop of liquid.

- The pack is made transparent to let the user to see how much more product is left inside.

- This concept can be used for any liquid product for any end use sector.

# Block Game for Kids

**CHANGING TRENDS IN COSMETICS PACKAGING**

## Manufacturer/Designer

**Developed by Fontos
Graphic Design Studio**

- The packaging is targeted for kid's products like shampoo, massage oil, hair oil, body lotions, etc

- The specific shape gives ease in stacking.

- One another advantage — the packs can be used as "blocks" for the kids to play with.

- Bright colours are in conjugation to attract kids

## Manufacturer/Designer

**Developed by Paul Capili**
**(for fictional brand Super Skin)**
**p.capili@gmail.com**

- The idea is taken from a well known comic book heavyweight character-Marvel.

- The design with a tagline that reads "even superheroes need protection from the sun's harmful rays." gives attractiveness to the pack

- The pack is very attractive being in the shape of capsule.

# Carton Look-Alike

## Manufacturer/Designer

**Carolin Boström, Broby Grafiska**
**http://www.recreatepackaging.com**

- The material used is a board, of-course it doesn't look like it.

- Advantage is the amazing shape and light weight compared to rigid plastic packaging

- The material provides lightness to the otherwise looking heavy bottle / pack

# Multipurpose Packaging

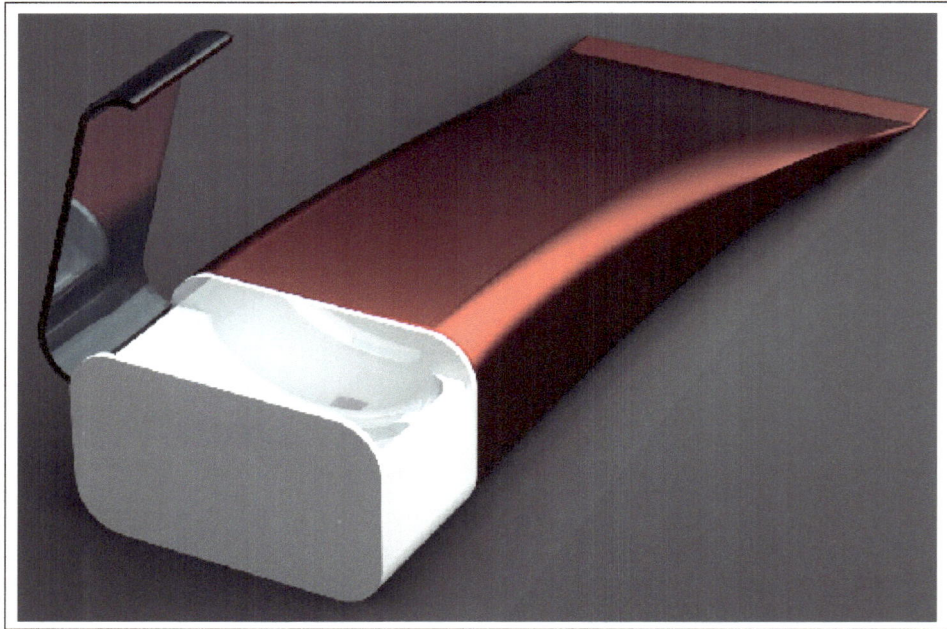

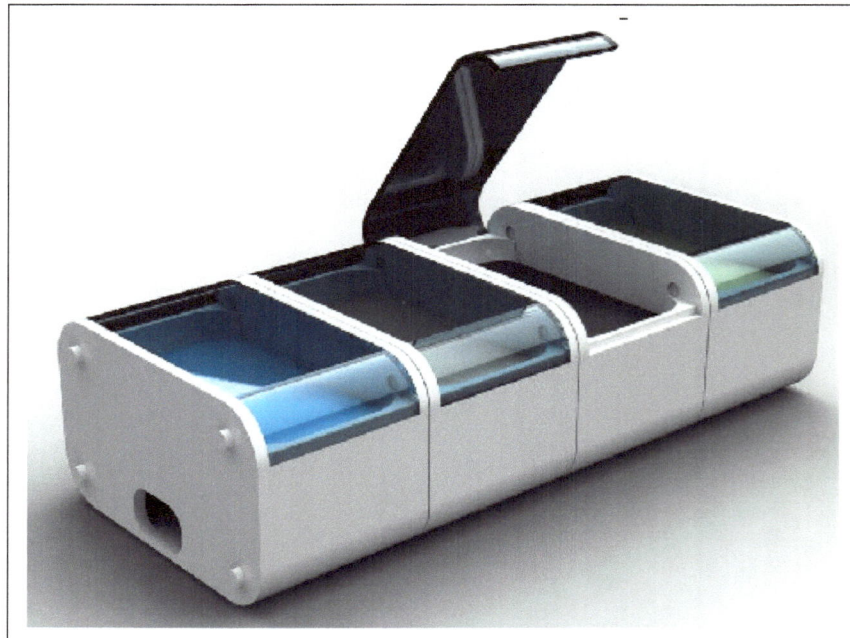

**CHANGING TRENDS IN COSMETICS PACKAGING**

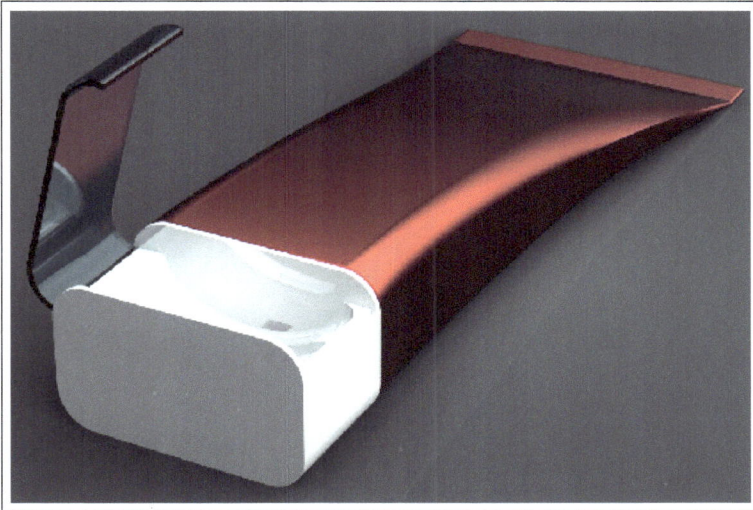

## Manufacturer/Designer

**Contact person – Georges
Koussouros (freelance inventor)
G.K@PROinvention.com
http://www.PROinvention.com**

- The product is dispensed in the closure (in form of jar)

- Once the product is completely used, jar can be detached from the

- The jar can be used for filling / mixing purpose as well.

- Dry products like face powder, eye shadows can be filled in the same

- Also jars can be collected and joined for building personal sets

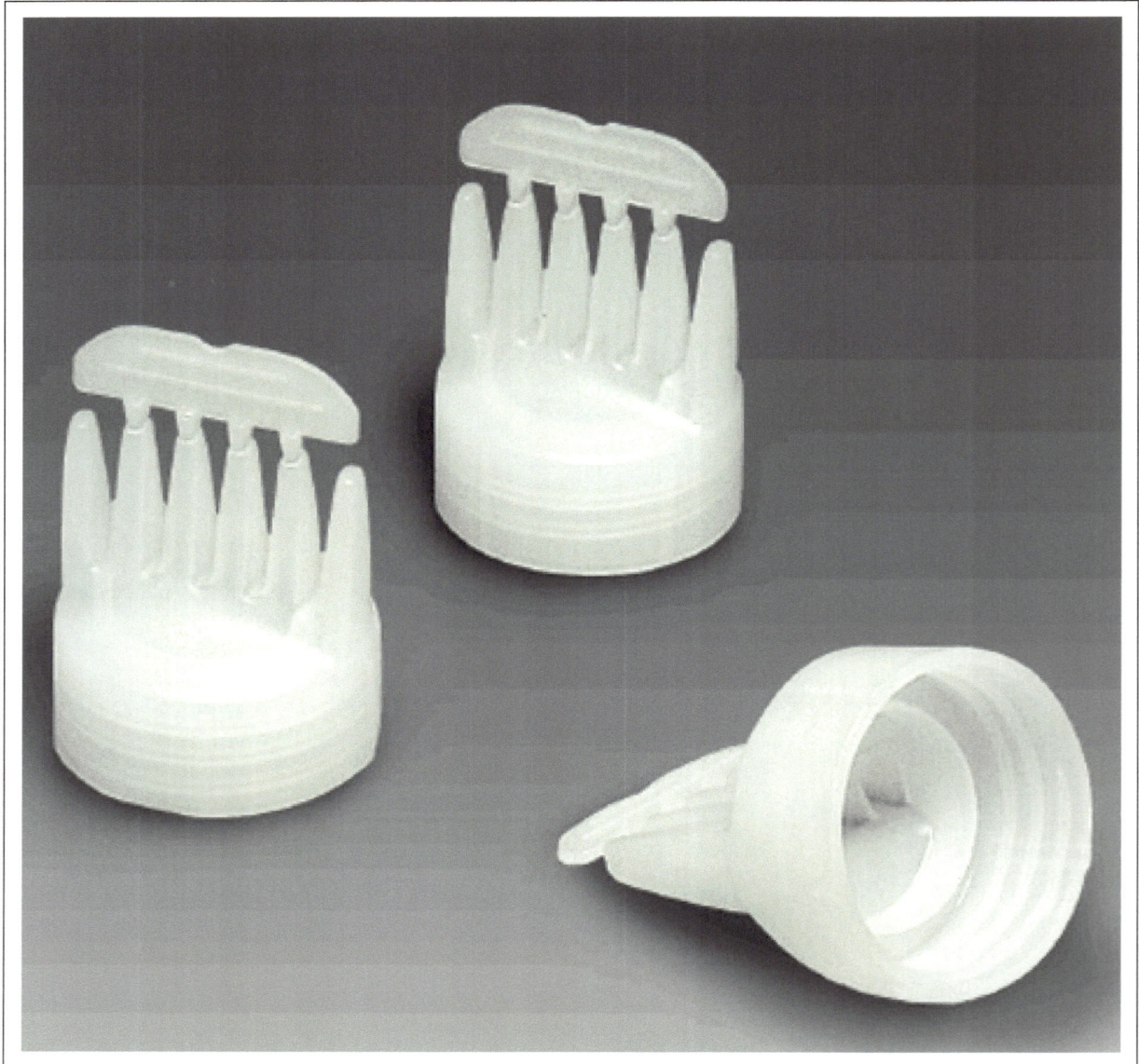

## Manufacturer/Designer

**Manufactured by Co. Plast.**
**Italy 031860077-031860135**
**info@coplast.it "www.coplast.it**

- An innovative applicator that evenly distributes the product and ensures a perfect and regular application.

- The applicator parts hair, controls release and evenly distributes the colour rights down to the roots

- The round-ended branches allow a soft application and prevents any skin irritation with minimum deposits on the scalp.

- Entirely made of PE, the applicator has a screw-on neck

# GREEN PACKAGING

**CHANGING TRENDS IN COSMETICS PACKAGING**

## Manufacturer/Designer

**Designed by Jung Hyun Jee**
**More details not available**

- The new design for packaging of soap / shampoo provide an eco friendly option

- The material used to make outer pouch can be given various effects (matt, smooth, rough) so as to give the feel of the product without even seeing the pack

- In this case material used is corn starch

- A suction cup is given on back panel to allow sticking of the pack with the wall

**CHANGING TRENDS IN COSMETICS PACKAGING**

## Manufacturer/Designer

**Developed by Albea Group**
**www.albea-group.com**
**In use for E.Leclerc**
**(one of the top French Retailers)**

- The tubes combine printing quality, innovation and eco-design.

- The white, pearlized tubes have several specificities:

- Their sleeve is 60% made from recycled milk bottles (PCR)

- The second tube has an ultra-light Access DeniedTM cap (with 40% less material and tear off band that gives temper proof feature))

- The first tube is a PrécitubeTMwith an airless pump that ensures maximum bulk protection and delivers just the right amount.

- Both products have been certified with the Ecocert label and the Cosmebio charter.

**CHANGING TRENDS IN COSMETICS PACKAGING**

## Manufacturer/Designer

**Developed by Yonwoo Korea**
**http://www.quadpack.net**

- "Paper Blow" is eco-friendly airless pack that conforms to the four Rs- Refill, Reuse, Reduce, Recycle.

- It features an overcap and outer bottle made of recycled PCR cardboard. These are fitted around a tube-like PE pouch attached to an airless pump.

- These are blow inserted into the cardboard bottle using surface friction, for an extra-tight fit.

- The PCR outer sheath means that no secondary packaging is necessary, reducing the overall materials used.

- Paper Blow is refillable eco friendly option for airless pump packs

## Manufacturer/Designer

**Developed by Braskem**
**http://www.quadpack.net**

- The new range of P&G Pantene shampoo uses packaging bottle made of renewable resource.

- The white bottles are made from a sugar cane based plastic and are 100% recyclable

- Till date HD/LL are known to have been developed by renewable feedstocks

- This LDPE is expected to be commercialy available by Jan'14

- As per study "More than half, 59%, of shoppers state that seeing environmental claims on packaging positively impacts their behavior to either buy more of the brands they usually do or switch to others."

**CHANGING TRENDS IN COSMETICS PACKAGING**

# TECHNOLOGY

© Twistub Ltd 2011

**CHANGING TRENDS IN COSMETICS PACKAGING**

## Manufacturer/Designer

**Stephen Eldred (Co-Founder)**
**United Kingdom**
**+44 (0) 121 212 5942**
**http://www.twistub.co.uk**
**twistub@live.com**

- Refill cosmetic packaging
- Incorporated nozzle that is suitable for different viscosity products
- Consist of dispenser (to be twisted for dispensing) and refill pack (can be replaced with new)
- Cost saving since purchase include buying new refill only and not dispenser

## How it Works

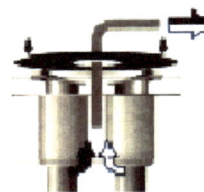

**50/50**

When the dial is aligned in the "middle" setting, the actualor disc will depress both pistons equally. Equal amounts of product will be dispensed from both pumps.

**30/70**

It between settings, e.g. 30/70, will result in product delivery in the ratios indicated.

## Manufacturer/Designer

### Manufactured by Revlon

- A "smart bottle" that enables customers to personalize the foundation to meet their skin tone.

- The foundation comes in a two-chamber bottle with a pump and a rotating dial.

- Different amounts of foundation will be dispensed from each chamber according to the number customers dial.

- The system is named Versadial™ from Versadial, Inc. and two related firms, Jarden Plastics and Yorker

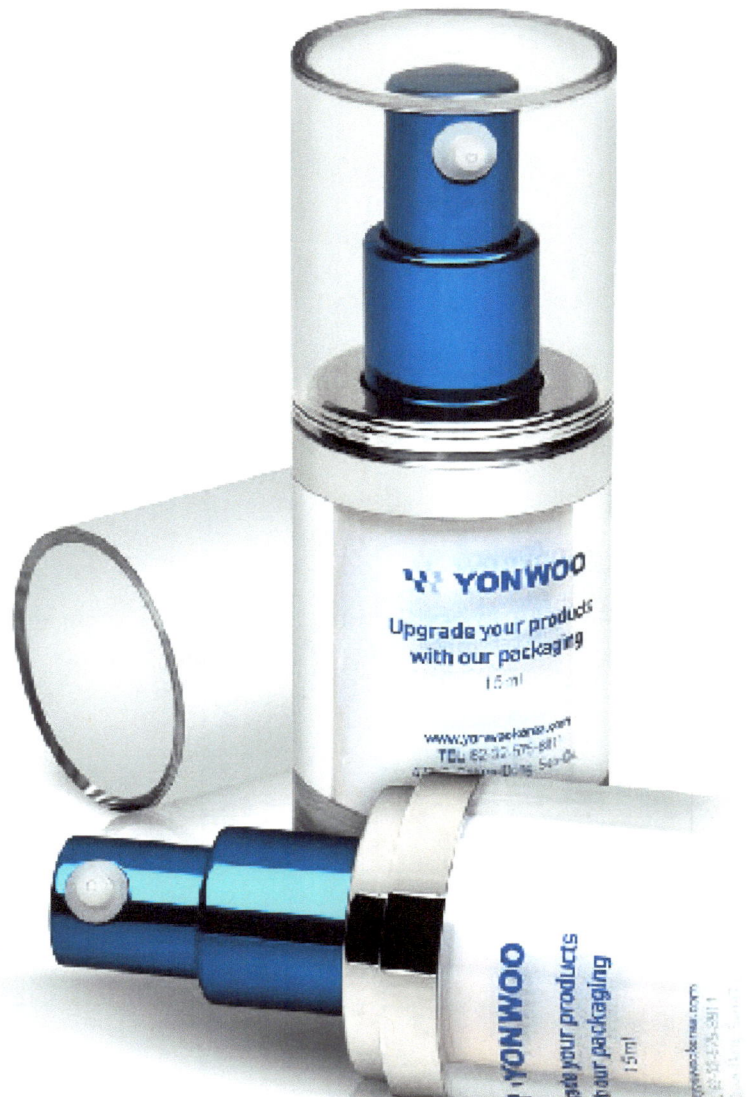

**CHANGING TRENDS IN COSMETICS PACKAGING**

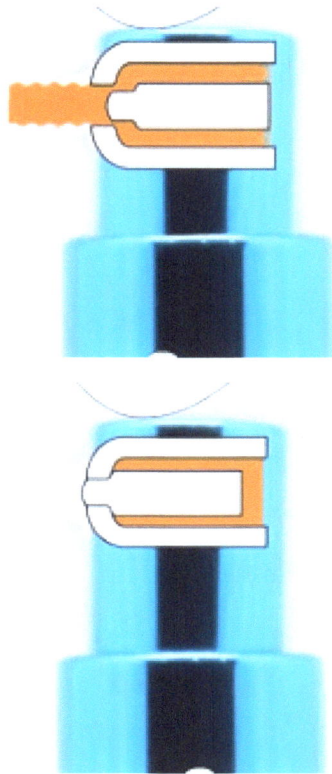

## Manufacturer/Designer

**Developed by Yonwoo Korea**
**http://www.quadpack.net**

- No matter how effective an airless container is, there is usually a degree of air contact inside the actuator.

- This technology prevents air re-entry and bulk exit through the actuator of an airless container.

- This ensures optimal protection and prevents oxidation of the bulk trapped in the nozzle.

- By not letting the product dry out at this end, it also avoids clogging, particularly when the pack is left on the shelf for any length of time.

- Functionality is deceptively simple. The pressure of the finger on the actuator opens the orifice and pushes the bulk outward. As soon as pressure on the actuator is released, the orifice immediately closes, isolating the product inside the nozzle from the outside.

**CHANGING TRENDS IN COSMETICS PACKAGING**

**CHANGING TRENDS IN COSMETICS PACKAGING**

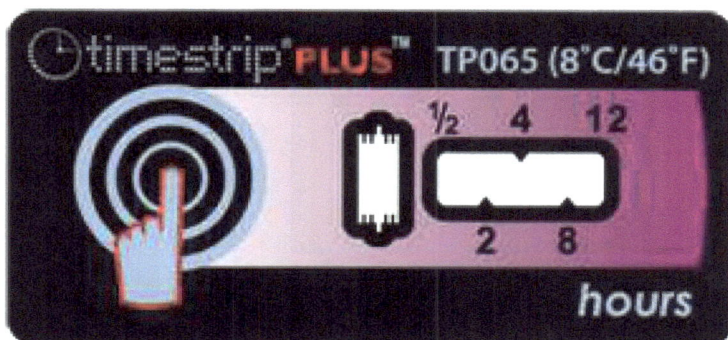

## Manufacturer/Designer

**Developed by Timestrip**
**+1 516-441-0133 / 44 (0) 8450 944 123**
**http://www.timestrip.com**
**First time used by Cargo Cosmetics"info@cargocosmetics.com**

- On opening the lip gloss for the first time, a consumer can insert the strip into the lip-gloss cap which activates the technology.

- The window begins to turn red and when, after nine months, the entire window is red, the consumer knows the product has reached its recommended expiry date

- Can be used for any cosmetic where there is PAO (period after opening) symbol

- As per European Cosmetics Directive, product with shelf life >30 months excluding single use products must display PAO symbol on both primary and secondary packs

- The technology can be used for cold storage check, blood banks, pharma, food, cosmetics

- Both time and temperature monitoring options are possible

**CHANGING TRENDS IN COSMETICS PACKAGING**

## CompelAroma TE®

Allows brand owners to protect product integrity with tamper seals while allowing consumers to sample the aroma of the product, connecting the consumer with the brand and driving purchase decisions.

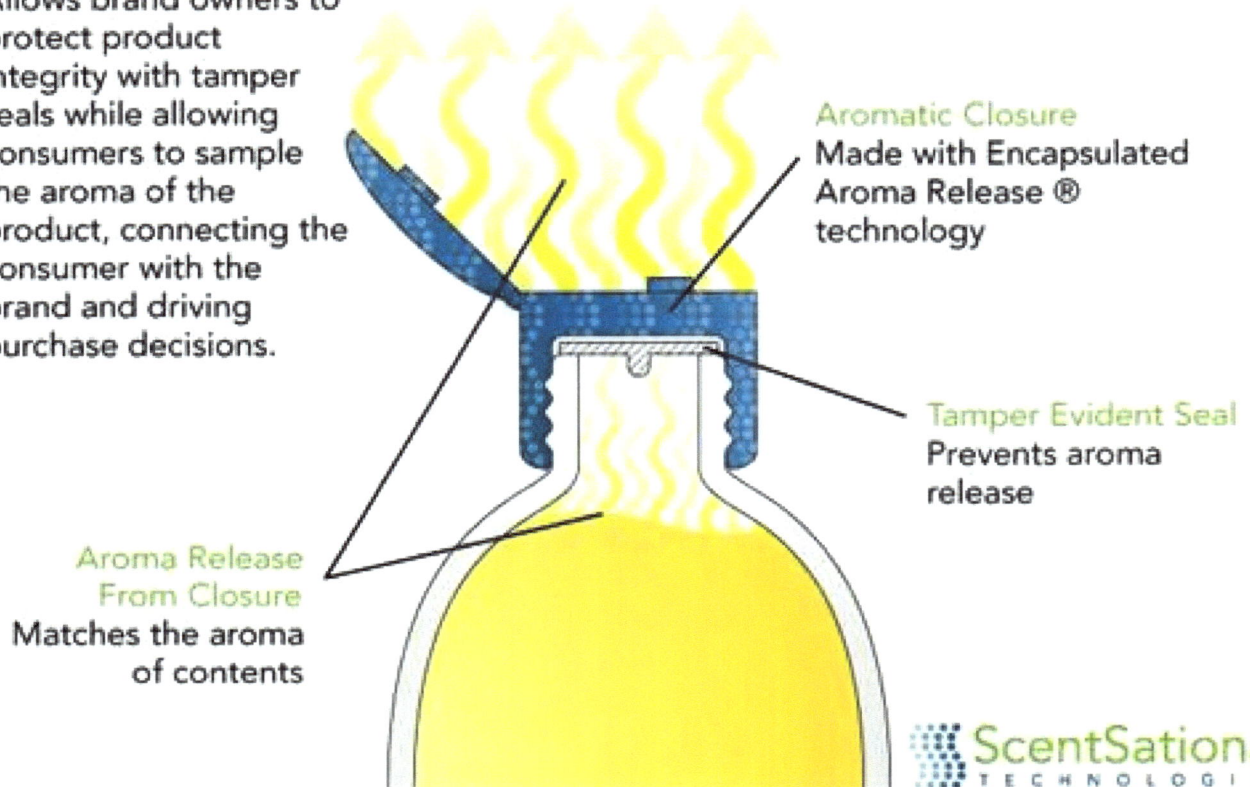

**Aromatic Closure**
Made with Encapsulated Aroma Release ® technology

**Tamper Evident Seal**
Prevents aroma release

**Aroma Release From Closure**
Matches the aroma of contents

ScentSational
TECHNOLOGIES

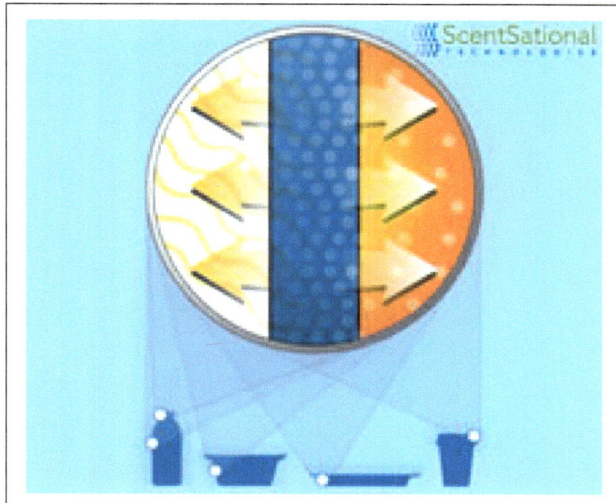

## Manufacturer/Designer

**ScentSational Technologies**
**Barry M. Edelstein (President)**
**BME@Scentt.com**
**http://scentsationaltechnologies.com**

A new, technology that will allow consumers to experience the aroma of the package contents without compromising any tamper evident systems in place.

The CompelAroma TE provides brands that are moving to tamper evident packaging the ability to reduce or eliminate reduction in sales that might result when consumers can no longer experience the aroma of their product in-store.

# PROMOTIONAL

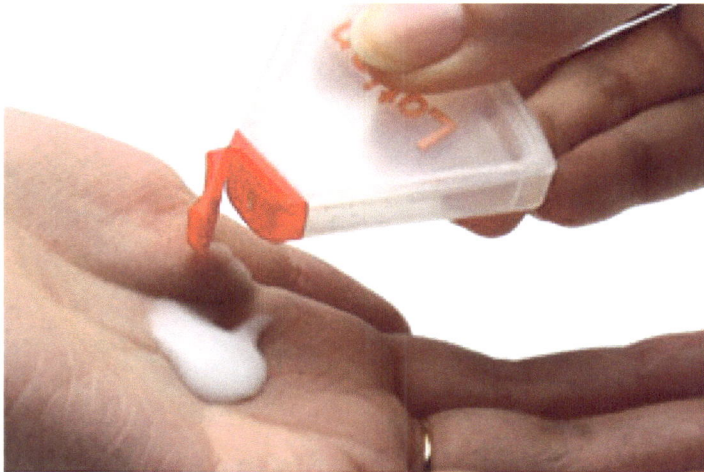

## Manufacturer/Designer

**Designed by Jin Chang-soo (further details not available) http://www.yankodesign. com/2010/08/04/ good-things-in-small-packages-2**

- This is a very creative example for a promo pack wherein small portions of different cosmetics products can be filled.

- Useful 'on-the-go' kit

- Can be refilled n number of times

- It has secure lock system that ensures no leaks, so no chances of shampoo or lotions mixing with each other or leaking on clothes

- Also this can be wall mounted for and allow multiple brands available at one place.

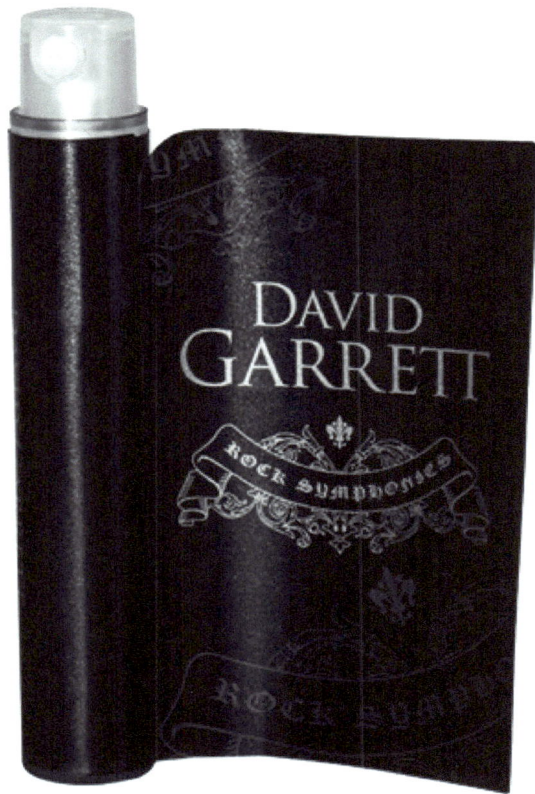

## Manufacturer/Designer

**Flacopac**
**Germany + 49 (0) 71 53 83 85 0**
**http://www.flacopac.com**

- The product – 'Flacoflag' allows the product itself to be flagged with the promotional message.

- It's a modern and dynamic way to put company's message right in front of the consumer

- Simple yet attractive idea to promote your product on the shelf

- Gives POP effect as well

## Manufacturer/Designer

**RLC packaging group**
**+49 511 164 99 - 0**
**rlc@rlc-packaging.com**
**http://www.rlc-packaging.com**

- Tubes lack an impact at the point-of-sales.

- The new carton tube packaging gives an innovative solution

- "Own display for every tube"

- The contour of the carton follows the shape of the tube like a second skin.

- The tube is attached at the cut-out at the front.

- The four-cornered standing surface enables a striking presentation on the shelf

**CHANGING TRENDS IN COSMETICS PACKAGING**

## Manufacturer/Designer

**Designed by Clondalkin Group**
**http://www.clondalkingroup.com/**
**Paperboard from Iggesund Paper board**
**http://www.iggesund.com/**
**Developed for Nivea**

- The design is an innovative, hexagonal carton

- This variant in gift packaging can be opened on both sides and has also passed all tests in practice.

- A cost-effective, yet singular design which allows brand to stand out in stores during a highly competitive season.

- The pack is an eye-catcher on the shelves and generates interest.

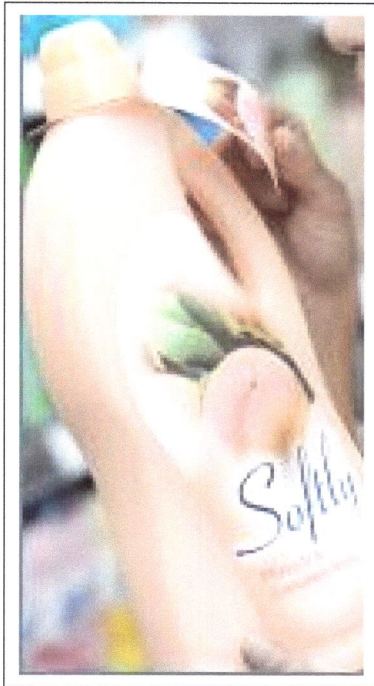

## Manufacturer/Designer

**Scent Printing Solutions**
**Tim Schwier (Follmann & Co)**
**+49 571 / 93 39 - 273**
**tim.schwier@scent-printing solutions.de**
**www.scent-printing-solutions.de**

- In retails stores battle for the customer's attention is particularly tough.

- The choice in the tiniest space is enormous and the rival product is just a few inches away, which makes it easy to compare prices. So how do we make up our minds?

- Fragrance is an effective instrument for boosting the buyer's confidence

- Example - fabric softener, the fragrance allows customer there and then to experience immediately the pleasure of fresh washed clothes, an incentive to pick up the product.

- Can be given through bottle tag, flag card, or through scent printing on the mono-carton

**CHANGING TRENDS IN COSMETICS PACKAGING**

# COSMETIC POWER

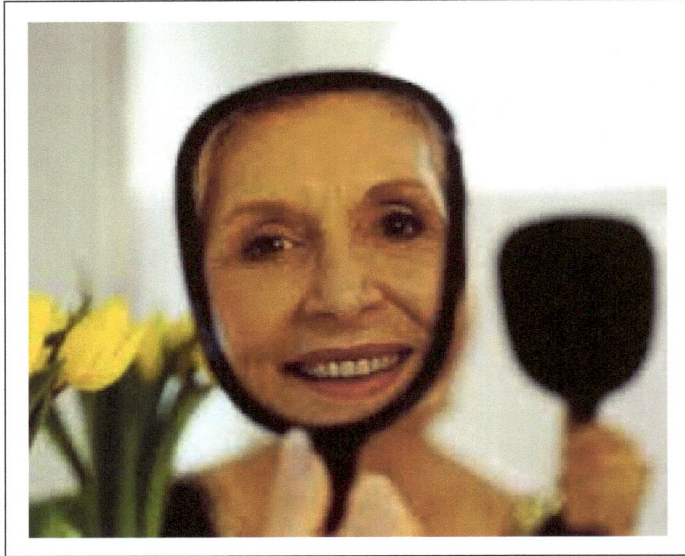

"Nature gives you the face
you have at twenty;
It is up to you to merit the face
you have at fifty."

www.ingramcontent.com/pod-product-compliance
Lightning Source LLC
Chambersburg PA
CBHW041729210326
41598CB00008B/825

*9788192792019*